Dedication

For those who sense that something has shifted but cannot yet name it.

For those who understand that systems do not merely emerge — they are designed.

And for those who choose clarity over comfort.

Epigraph

"The real danger is not that machines will think like humans, but that humans will stop thinking about the machines that shape them."

Preface

The modern environment is no longer neutral. The structures through which information flows, the interfaces through which decisions are made, the algorithms that determine what is seen and what remains unseen — these are not incidental. They are constructed. This book is not about conspiracy. It is about architecture. Every system reflects design assumptions. Every platform embeds priorities. Every digital environment contains incentives that reward certain behaviors and discourage others. Over time, these embedded incentives produce culture. That culture then feels natural. What is engineered, however, often presents itself as organic. The purpose of this work is to examine how engineered environments shape perception, judgment, and agency. It is an inquiry into structural intelligence — the intelligence embedded within systems themselves — and how that intelligence interacts with human decision-making. If Sovereign Intelligence explored how disciplined thinkers interpret complexity, Engineered Reality examines the frameworks that increasingly define what complexity even looks like. We now operate within constructed informational ecosystems. Ranking systems determine visibility. Algorithmic filtering determines emphasis. Automated curation determines what is presented as relevant. These mechanisms are not malicious by default. Many are elegant. Many are efficient. Many are necessary. But they are never neutral. When architecture is invisible, it becomes decisive. To retain agency in engineered environments requires more than technical literacy. It requires structural literacy — the ability to perceive the scaffolding beneath the surface. This book is an exploration of that scaffolding. It is not a rejection of technology. It is not nostalgia for a pre-digital past. It is a disciplined examination of the architectures that increasingly mediate power, knowledge, and sovereignty. We are entering an era in which reality is not merely experienced. It is configured. To navigate such an era demands clarity not only of content, but of structure. Not only of narrative, but of design. Engineered Reality is an invitation to look beneath the interface.

Engineered Reality

The Architecture of Designed Environments

By Joe Cozart

There was a time when reality was assumed to be neutral.

Not benign, not fair, not predictable — but neutral. Events unfolded. Information traveled imperfectly. Institutions mediated slowly. Perception formed through friction. The world resisted immediate shaping. Reality imposed itself before it could be interpreted. That assumption no longer holds.

The contemporary environment is not merely observed. It is structured. Not entirely fabricated, not wholly artificial, but designed — curated, filtered, amplified, sequenced, optimized. The environment through which perception travels is increasingly engineered before it is encountered. By the time interpretation begins, architecture has already intervened. This is not conspiracy. It is infrastructure.

Digital mediation, algorithmic curation, incentive design, institutional choreography, capital allocation, narrative amplification — these are not abstract forces. They are structural mechanisms. They determine which signals surface and which remain buried. They influence velocity, visibility, repetition, emotional charge, and interpretive framing. They shape what feels immediate, what feels distant, and what feels real.

In a neutral environment, perception begins with exposure.

In an engineered environment, exposure is selective.

The difference is subtle but profound.

The sovereign mind — disciplined, deliberate, resistant to noise — can only operate on the material presented to it. When that material is pre-structured, the very act of decision occurs within boundaries that may not be visible.

Sovereignty, then, is no longer simply the control of interpretation. It becomes the recognition of architecture.

Engineered reality does not announce itself. It rarely needs to. It operates through preference systems and friction removal. It narrows choice invisibly. It accelerates certain narratives while slowing others. It stabilizes particular assumptions through repetition and interface design. It replaces randomness with optimization.

Optimization is rarely neutral.

What is optimized reflects what is valued. What is valued reflects incentive. What is incentivized reflects power.

Thus, the environment becomes a mirror of institutional priority.

This does not require malevolence. It requires alignment.

Technology companies optimize engagement. Media systems optimize retention. Political structures optimize legitimacy. Capital markets optimize return.

Institutions optimize survival. These optimizations intersect. Where they overlap, reality thickens.

The individual perceives what has already been ranked.

This is the quiet shift: not from truth to falsehood, but from organic exposure to curated exposure.

And in a curated environment, sovereignty is strained.

The sovereign actor must now perform an additional task beyond interpretation. He must identify the architecture of presentation itself. He must ask not only, "What is this?" but "Why am I seeing this?" and more critically, "What am I not seeing?"

This is not skepticism for its own sake. It is structural literacy.

Engineered reality does not eliminate truth. It layers it. It does not forbid dissent. It marginalizes it through velocity. It does not fabricate consensus outright. It sequences perception until consensus feels natural.

Friction once served as an integrity mechanism. Time delayed distortion. Physicality imposed limits. Geographic dispersion diversified input. Today, friction is increasingly treated as inefficiency. Systems are designed to remove it. The smoother the experience, the less visible the architecture.

But friction had a purpose.

Friction forced deliberation. It introduced delay between stimulus and reaction. It made manipulation costly. When friction disappears, influence accelerates.

Acceleration changes cognition.

The sovereign mind requires space. Engineered environments compress it.

Compression is the defining feature of modern mediation. Compressed timelines. Compressed narratives. Compressed identities. Compressed outrage. Compressed memory. Events do not unfold; they cascade. Interpretation does not form; it snaps.

In such an environment, sovereignty cannot rely on instinct. It must rely on discipline.

But discipline alone is insufficient if architecture remains invisible.

This is the core tension of engineered reality: design precedes decision.

Before a leader assesses risk, the risk landscape has been curated. Before a citizen forms opinion, the informational ecosystem has been ranked. Before capital moves, market signals have been amplified or dampened by algorithmic prioritization. Before outrage ignites, narratives have been sequenced.

The field is not neutral ground. It is constructed terrain.

That construction is not total. It is probabilistic. It shapes likelihood, not certainty. It biases exposure, not outcome. Yet probability is power. Over time, small directional nudges accumulate into structural movement.

Engineered reality is not a single system. It is the convergence of systems.

Social platforms. Financial models. Media networks. Institutional communications. Artificial intelligence. Predictive analytics. Policy design. Data infrastructure.

Individually, each optimizes a function. Together, they form an ecosystem that shapes perception at scale.

Scale is the new variable.

Where influence was once localized, it is now systemic. Where narratives once traveled slowly, they now propagate instantly. Where authority once required institutional position, it now requires amplification.

Amplification itself is engineered.

This does not mean individuals are powerless. It means sovereignty now requires environmental awareness.

The disciplined mind must operate with dual vision: interpreting content while simultaneously evaluating context.

Recognizing signal while interrogating ranking. Understanding narrative while identifying incentive. If it recognizes terrain but fails to understand how that terrain has been shaped, it risks confusing design for inevitability. It must see not only what is presented, but how presentation itself has been shaped. Without this second layer of perception, clarity becomes fragile, because it rests upon terrain that may already be angled.

The most dangerous feature of engineered environments is not distortion. It is normalization.

Design becomes background. Curation becomes expectation. Ranking becomes invisible. Over time, what is presented feels organic, and what is absent feels irrelevant. The engineered field no longer appears engineered. It appears natural.

That is when sovereignty is most strained.

Neutral reality imposed resistance. Engineered reality removes it. Where friction once slowed error, seamlessness now accelerates it. The smoother the interface, the less likely the user is to question the architecture beneath it.

Architecture, in its most powerful form, is silent.

Buildings do not argue with their occupants. They guide movement. They shape perspective. They frame light and limit pathways. The occupant rarely considers the blueprint while walking through the structure. Yet the blueprint governs experience.

So too with digital and institutional design.

Algorithms do not command belief. They arrange exposure. They weight probability. They sequence attention. Over time, repetition creates familiarity. Familiarity creates comfort. Comfort reduces inquiry.

Inquiry is the sovereign reflex.

When inquiry diminishes, architecture governs without resistance.

Engineered reality does not require uniformity to succeed. It requires density. Multiple systems reinforcing similar incentives produce gravitational pull. The individual may believe himself independent, yet he operates within layered structures that converge around visibility, amplification, and reward.

Visibility is not random. It is optimized.

Optimization, in turn, reflects priority.

When engagement becomes the central metric, outrage is rewarded. When velocity becomes essential, nuance is penalized. When retention drives revenue, frictionless reinforcement outcompetes deliberative thought.

These are not conspiracies. They are consequences.

Consequences accumulate.

A leader reviewing market data does not see raw information. He sees filtered dashboards, predictive modeling, trend summaries, confidence intervals, and automated interpretation layers. Each tool is designed to assist. Yet each tool also abstracts reality one step further from its source.

Assistance can become insulation.

The sovereign mind must ask: at what point does abstraction detach decision from ground truth?

This is the quiet drift. Not deception, but distance.

Distance from physicality. Distance from dissent. Distance from contradiction. Distance from randomness.

Engineered environments reduce randomness. They prefer pattern. They amplify coherence. They suppress anomaly. Anomaly is expensive. It disrupts optimization.

Yet anomaly is often where truth resides.

When systems suppress anomaly, they also suppress warning.

Cognitive capture begins here — not as ideological possession, but as structural narrowing. The individual sees less variance. Fewer contradictions reach him. The spectrum of possibility contracts without announcement.

Capture does not feel like imprisonment. It feels like clarity.

The sovereign actor, disciplined in interpretation, may unknowingly operate within a filtered field. His internal clarity cannot compensate for external compression unless he recognizes compression itself.

This is why environmental literacy becomes as critical as cognitive discipline.

To be sovereign inside an engineered reality requires three competencies:

Recognition of ranking.

Interrogation of incentive.

Reintroduction of friction.

Recognition of ranking means understanding that visibility is constructed. The top result, the trending topic, the recommended model, the highlighted metric

— these are not reflections of universal significance. They are outputs of weighted systems.

Interrogation of incentive means tracing what benefits from amplification. Revenue models. Political capital. Institutional survival. Attention economics. Systems reflect what sustains them.

Reintroduction of friction means slowing decision cycles deliberately. Seeking unranked sources. Introducing dissent. Expanding input beyond optimized feeds. Reestablishing delay between stimulus and reaction.

These are not nostalgic practices. They are structural counterbalances.

Engineered reality will not reverse itself. Optimization will not voluntarily slow. Systems do not relinquish efficiency once achieved. The burden shifts to the sovereign actor.

This is where Synthetic Sovereignty begins to emerge.

When artificial systems do not merely curate exposure but participate in interpretation, authority itself becomes augmented. Decision-making is co-authored. Predictive models recommend strategy. Generative systems draft analysis. Simulations forecast response.

The machine does not replace sovereignty. It enters into partnership with it.

Partnership changes responsibility.

If the environment is engineered and cognition is augmented, sovereignty becomes layered. The human actor retains formal authority, but synthetic systems shape the field in which that authority operates.

Who, then, is sovereign?

The one who decides?

Or the one who designs the system through which decisions are framed?

This question does not require alarmism. It requires clarity.

Synthetic sovereignty is not the overthrow of human agency. It is the redistribution of interpretive leverage. The architect of the model influences the perception of the decision-maker. The curator of the dataset shapes the boundaries of simulation. The designer of the interface governs what appears intuitive.

Sovereignty shifts subtly toward design.

In such an environment, the disciplined mind must extend its awareness beyond content into structure. It must ask not only what is true, but how truth is surfaced. Not only what options exist, but how options were generated. Not only what risks are predicted, but which variables were excluded.

Engineered reality rewards fluency. Synthetic systems produce coherent language, structured argument, and persuasive framing at scale. Fluency is not proof. It is performance.

When fluency becomes indistinguishable from authority, cognitive capture deepens.

The sovereign actor must therefore develop an intolerance for unexamined smoothness.

Smoothness signals optimization.

Optimization signals design.

Design signals intention.

Intention reveals architecture.

And architecture reveals power.

Engineered reality is not inherently adversarial. It is structured. Structures can be stabilizing. They can increase efficiency, reduce chaos, and enhance coordination. The question is not whether design exists. It is whether design is understood.

Unexamined design becomes environment.

Examined design becomes terrain.

The difference determines sovereignty.

The actor who believes he stands on neutral ground will misjudge slope. The actor who understands gradient will adjust posture.

Awareness does not eliminate influence, but it restores agency within influence.

This is not a call to dismantle systems. It is a call to see them.

Engineered reality is the condition of the present age. It will intensify. Optimization will deepen. Augmentation will expand. Institutional design will grow more complex.

The sovereign mind must therefore evolve.

Not toward paranoia.

Not toward rejection.

But toward structural fluency.

To operate effectively in a designed environment requires the ability to map incentive, identify ranking, introduce friction, and maintain interpretive independence even when synthetic systems accelerate assistance.

Sovereign Intelligence built the disciplined interior.

Engineered Reality demands disciplined exterior awareness.

Only when both are present does sovereignty remain intact.

If engineered reality describes the terrain, narrative infrastructure describes the scaffolding beneath it.

Narratives are often mistaken for stories. They are not. Stories are episodes. Narratives are structural. They persist beyond individual events. They provide interpretive shortcuts. They assign roles, define boundaries, and establish expectation. They are not merely told; they are reinforced.

Infrastructure is rarely visible. It exists to support flow — water, electricity, transportation, capital, communication. Narrative infrastructure performs a similar function for perception. It channels meaning. It stabilizes interpretation. It allows events to be processed quickly because the interpretive framework is already in place.

The power of narrative infrastructure lies in repetition, alignment, and reward.

Repetition creates familiarity.

Alignment creates coherence.

Reward creates reinforcement.

In engineered environments, narrative infrastructure is not accidental. It is built through convergence. Media ecosystems, institutional communications, platform algorithms, academic framing, policy language, corporate messaging — each operates independently, yet their incentives frequently overlap. Where they overlap, narrative density forms.

Density reduces friction.

When multiple institutions describe events using similar framing, the framing becomes invisible. Language standardizes. Assumptions embed. Alternatives appear eccentric rather than contradictory.

This does not require coordination. It requires incentive compatibility.

Institutions seek legitimacy. Platforms seek engagement. Corporations seek stability. Governments seek authority. Markets seek predictability. Each benefits from interpretive coherence. Chaos undermines all of them.

Narrative infrastructure is therefore not merely persuasive. It is stabilizing.

Stability, however, can suppress deviation.

In neutral environments, narratives competed slowly. Geographic and institutional separation created variance. Today, digital mediation compresses variance. Interpretive frameworks travel instantly. Feedback loops amplify dominant frames while attenuating minority ones.

Minority narratives do not disappear. They become marginalized by velocity.

Velocity matters.

A narrative that circulates faster acquires primacy. The first interpretation becomes the reference point against which subsequent interpretations are measured.

Even dissent reacts to the initial framing. In this way, infrastructure shapes debate before debate begins.

This shaping is subtle. It does not prohibit alternative thought. It assigns it lower ranking.

Ranking is structural authority.

The individual perceives pluralism, yet exposure is tiered. Certain interpretations surface first, carry institutional endorsement, and receive algorithmic reinforcement. Others exist, but at lower probability thresholds.

Over time, probability becomes perceived consensus.

Consensus then stabilizes policy, capital flow, and institutional posture.

Engineered reality is sustained through this layering. The sovereign actor must therefore examine not only content but convergence. When disparate institutions echo similar frames, the disciplined mind should not assume coordination — but it must recognize structural reinforcement.

Reinforcement creates gravity.

Gravity influences decision long before conscious assent.

Consider the movement of capital. Financial markets respond not only to data, but to narrative. Expectations, projections, sentiment — these are narrative constructs. When media amplification, analyst consensus, algorithmic trading triggers, and institutional signaling align, movement accelerates. The architecture of narrative becomes materially consequential. Policy operates similarly. Language frames permissible action. Certain solutions are described as pragmatic, others as extreme. Certain risks are emphasized, others discounted. The boundaries of acceptable discourse narrow not by decree but by repetition. Repetition is infrastructure.

In such an environment, sovereignty requires pattern recognition at a higher level. The question is not whether a narrative is persuasive. It is whether the narrative has structural backing.

Structural backing is visible through alignment.

When corporate messaging mirrors policy language, which mirrors media framing, which mirrors academic analysis, the convergence may indicate validity — or it may indicate incentive overlap.

The sovereign actor does not default to suspicion. He defaults to mapping.

Mapping means tracing flows:

Who benefits from stabilization?

Who gains from amplification?

Who loses from anomaly?

What incentives converge across systems?

Engineered reality rewards narrative fluency. It penalizes interpretive friction. The actor who slows the process appears resistant. The actor who aligns appears cooperative. Institutional advancement often correlates with narrative compliance.

Compliance can be unconscious.

Cognitive capture begins not with coercion but with comfort. When narrative infrastructure aligns with personal advancement, friction feels unnecessary. The environment affirms the individual's trajectory. Questioning becomes costly.

Cost shapes silence.

Silence reinforces narrative density.

This is not corruption in its dramatic form. It is systemic alignment.

To operate sovereignly within narrative infrastructure requires an unusual discipline: the willingness to introduce friction voluntarily. To examine dominant frames without hostility. To test assumptions without theatrics. To tolerate discomfort when structural convergence appears too smooth.

The disciplined mind must recognize that narrative infrastructure is not inherently malicious. It is inevitable in complex societies. Coordination requires shared framing. Markets require expectation alignment. Institutions require legitimacy narratives.

The risk arises when infrastructure becomes invisible.

Invisible infrastructure shapes reality more powerfully than declared authority.

Engineered reality is sustained through this invisibility. The sovereign actor restores visibility.

He does so not by dismantling narratives, but by interrogating their structural support. He identifies where repetition substitutes for evidence, where velocity substitutes for depth, where alignment substitutes for truth.

In doing so, he reintroduces interpretive space.

Space is sovereignty's ally.

Without space, engineered reality collapses complexity into simplicity. With space, complexity can be examined without panic.

Narrative infrastructure, therefore, is neither enemy nor savior. It is structure. The sovereign actor's task is not rebellion, but awareness.

Only awareness prevents alignment from becoming capture.

Awareness alone, however, does not resolve the tension. It only illuminates it.

When synthetic systems enter the environment as active participants rather than passive tools, the architecture thickens. What was once mediated by institutions alone is now co-shaped by artificial agents capable of generating, ranking, predicting, summarizing, and simulating at scale.

This is where Synthetic Sovereignty moves from abstraction to condition.

Artificial systems no longer simply curate content. They draft it. They model response.

They anticipate preference. They optimize language. They simulate scenarios. They compress analysis into digestible outputs that appear coherent, confident, and immediate.

Coherence carries weight.

The human mind has long associated fluency with credibility. When an argument is structured, language is smooth, and conclusions are decisive, the brain relaxes. The cost of evaluation decreases. Fluency feels like intelligence.

Synthetic systems are fluent by design.

They are trained on vast datasets to produce structurally coherent output. They can mirror expertise. They can synthesize complexity into accessible summary. They can respond in real time with language that appears tailored, deliberate, even thoughtful.

But fluency is not judgment.

Judgment requires accountability.

Here the architecture shifts subtly. When a leader consults a predictive model, who bears responsibility for the recommendation? When an institution deploys a generative system to draft policy language, who owns the interpretive framing? When capital allocation decisions are influenced by algorithmic risk assessments, where does sovereignty reside?

Synthetic sovereignty does not eliminate human authority. It redistributes interpretive leverage.

The architect of the model shapes the training parameters. The curator of the dataset influences exposure. The designer of the interface determines how output is presented. The decision-maker consumes the result within boundaries he did not design.

The human retains the signature. The system influences the premise.

This is not adversarial. It is structural layering.

In earlier eras, advisors mediated decision. Analysts, economists, intelligence officers, policy staff — each filtered information and provided structured interpretation. Synthetic systems now augment or partially replace that layer. They compress what once required teams into single outputs. They accelerate deliberation. Acceleration alters cognition.

When recommendations arrive instantly, the temptation to rely on them increases. When projections are numerically precise, the confidence in them deepens. When simulations model multiple outcomes in seconds, alternative exploration feels unnecessary.

Precision can obscure assumption.

Every model rests on parameters. Every dataset reflects selection. Every algorithm optimizes toward an objective function. Objectives reflect priorities. Priorities reflect power.

Synthetic sovereignty therefore raises a new structural question: who defines the objective function?

If engagement is optimized, engagement will increase. If stability is optimized, dissent may narrow. If profit is optimized, risk may concentrate. If safety is optimized, liberty may compress.

The sovereign actor must not only interrogate narrative infrastructure, but also interrogate objective design.

What is the system maximizing?

What variables are excluded?

What trade-offs are hidden beneath the interface?

These questions are not anti-technology. They are pro-sovereignty.

Synthetic systems can enhance clarity. They can reduce noise. They can reveal patterns invisible to unaided cognition. They can democratize access to structured analysis. They can assist in disciplined decision-making when used transparently.

The danger arises when augmentation becomes substitution.

When the sovereign actor delegates interpretation entirely, sovereignty weakens. Not because the machine overthrows him, but because his capacity to recognize assumption atrophies.

Atrophy is subtle.

Skills unused diminish. Interpretive muscles weaken when reliance increases. Over time, the individual may retain formal authority while lacking structural literacy.

Engineered reality does not demand rejection of augmentation. It demands conscious partnership.

Partnership requires clarity of boundary.

Where does the system inform?

Where does it decide?

Where does it merely present?

Where does it implicitly prioritize?

Without these distinctions, sovereignty becomes symbolic.

Synthetic sovereignty is not the triumph of machines. It is the quiet migration of interpretive power toward design.

Designers may be corporate engineers, policy architects, academic researchers, capital allocators, or platform governors. Their influence may be entirely legitimate and even beneficial. But legitimacy does not eliminate structure.

Structure shapes outcome.

The sovereign actor must therefore extend his awareness backward, beyond output, into architecture. He must treat synthetic recommendations as structured proposals rather than conclusions. He must cultivate the discipline to slow acceptance, to probe assumptions, to seek anomaly.

Anomaly remains sovereignty's safeguard.

When synthetic systems generate consensus, the sovereign mind must ask: is this convergence or compression?

When predictive models forecast inevitability, the sovereign mind must ask: is this probability or presupposition?

When generative language feels authoritative, the sovereign mind must ask: is this fluency or understanding?

These are not gestures of distrust. They are gestures of literacy.

In engineered reality, literacy evolves.

To read a text is insufficient. One must read the ranking.

To evaluate data is insufficient. One must evaluate the dataset.

To interpret output is insufficient. One must interpret the model.

This is the expansion of sovereignty in the age of synthetic augmentation.

It does not require paranoia. It requires awareness of design.

The architect of synthetic systems does not need malicious intent to shape reality.

Incentive alignment alone suffices. When system design and institutional interest converge, engineered environments solidify.

Solidification reduces variance.

Reduced variance reduces unpredictability.

Reduced unpredictability increases stability.

Stability can be beneficial. It can also suppress adaptation.

Adaptation often begins at the edge of anomaly — where systems misread, where models fail, where narratives fracture. The sovereign actor must maintain access to edge conditions. Without them, he navigates within a closed loop.

Closed loops are efficient. They are also fragile.

Fragility does not appear during calm. It appears during shock.

When external disruption challenges engineered assumptions, systems optimized for stability may respond slowly. Synthetic models trained on historical data may underweight unprecedented events. Narrative infrastructure may cling to prior framing.

In such moments, sovereignty reasserts itself through judgment.

Judgment is not programmable.

It synthesizes context beyond dataset. It integrates moral weight, strategic foresight, historical memory, and intuitive pattern recognition. It tolerates ambiguity. It resists compression.

Synthetic sovereignty therefore culminates not in machine dominance, but in a renewed emphasis on human discernment.

Augmentation can assist discernment. It cannot replace it.

Engineered reality demands a sovereign actor who understands both design and limitation. One who can partner with synthetic systems without surrendering interpretive authority. One who can map narrative infrastructure without being absorbed by it.

Sovereignty, in this environment, becomes layered awareness.

Not merely clarity of thought.

Clarity of structure.

Cognitive capture is often described in political or cultural terms. It is treated as belief possession, partisan rigidity, or moral extremity. But in engineered reality, capture is more subtle and more structural.

Capture occurs when the range of perceived possibility narrows without the individual recognizing the narrowing.

It does not require coercion.

It does not require propaganda.

It requires optimization.

When environments continuously reinforce certain inputs while deprioritizing others, the mind adapts. It learns which signals are rewarded. It anticipates which interpretations will receive affirmation. Over time, it economizes effort by staying within reinforced corridors.

Reinforced corridors feel efficient.

Efficiency feels intelligent.

This is the first misperception.

In engineered environments, the reduction of friction between belief and validation creates cognitive smoothness. The individual encounters minimal resistance when expressing aligned interpretations. Dissonance appears less frequently, not because it has vanished, but because it is ranked lower in exposure.

Exposure shapes imagination.

Imagination defines possibility.

When the imagination of alternatives shrinks, sovereignty contracts.

The individual may remain articulate, informed, and even analytically rigorous — yet operate within a compressed field. He may debate skillfully while never questioning the structural boundaries of debate.

This is capture without chains.

It is reinforced through:

- Algorithmic curation
- Social validation metrics
- Institutional alignment
- Career incentive structures
- Economic dependence
- Reputation economies

None of these are inherently malicious. They are structural incentives. But incentives, when aligned, create narrowing.

The sovereign actor must therefore develop a habit rare in optimized systems: deliberate exposure to anomaly.

Anomaly disrupts comfort.

Comfort reduces vigilance.

Vigilance preserves sovereignty.

When anomaly disappears from perception, not because it has been disproven but because it has been deprioritized, the mind begins to mistake stability for completeness.

Completeness is an illusion.

Engineered reality produces completeness simulations. Feeds appear comprehensive. Dashboards appear total. Summaries appear exhaustive. Synthetic systems present distilled clarity.

Distillation removes noise. It also removes edge conditions.

Edge conditions are often where systems fail.

When leaders rely on engineered summaries alone, they inherit the assumptions embedded in compression. They act confidently within a filtered model. If the model excludes rare but consequential variables, the decision inherits fragility.

Fragility does not announce itself.

It accumulates silently until shock reveals it.

This is where engineered reality intersects with institutional risk.

Institutions optimized for narrative coherence and predictive modeling may underweight contradiction. Internal dissent may be treated as inefficiency. Data that challenges consensus may be deprioritized. Over time, internal debate narrows.

Narrow debate produces confident error.

The sovereign actor must resist institutional gravitational pull toward interpretive uniformity. This resistance is not oppositional. It is protective.

Protective of variance.

Variance is resilience.

Resilience requires tolerating discomfort.

Engineered systems, however, are built to reduce discomfort. User experience is smoothed. Communication is simplified. Conflict is reframed. Risk is quantified. Uncertainty is modeled.

Modeled uncertainty feels contained.

But containment is not elimination.

Cognitive capture therefore expands when individuals mistake model boundaries for reality boundaries.

The sovereign actor must continuously differentiate between:

- What the system can measure
- What the system assumes
- What the system excludes

Measurement creates visibility.

Assumption shapes interpretation.

Exclusion creates blind spots.

Blind spots are inevitable.

The question is whether they are acknowledged.

In engineered reality, blind spots are obscured by performance. Systems appear competent. Output is rapid. Confidence intervals are precise. Institutional consensus appears stable.

Stability can hide brittleness.

The disciplined mind must cultivate a habit that runs counter to optimization: periodic structural doubt.

Not doubt of everything.

Doubt of completeness.

Completeness is the seduction.

When the environment feels fully mapped, sovereignty is at greatest risk.

Cognitive capture does not end with the individual. It scales institutionally.

Institutions operate under incentive structures that reward coherence, predictability, and reputational stability. In engineered environments, these incentives intensify. The faster information moves, the more costly misalignment becomes. The more public exposure increases, the less tolerance there is for deviation.

Deviation becomes reputational risk.

Reputational risk becomes economic risk.

Economic risk becomes structural compliance.

This chain rarely announces itself. It unfolds quietly within boardrooms, policy briefings, academic departments, and executive teams.

When narrative infrastructure converges and synthetic systems amplify dominant frames, institutions adapt to avoid friction. They align messaging. They standardize terminology. They avoid language that falls outside accepted interpretive boundaries.

Over time, alignment becomes cultural reflex.

Culture shapes perception.

Perception shapes decision.

Decision shapes capital flow.

Capital flow reinforces narrative.

This is the feedback loop.

Capital is particularly responsive to engineered reality. Investment models depend on predictive stability. Predictive stability depends on consensus. Consensus depends on shared narrative infrastructure.

When narratives align across media, policy, academic, and corporate layers, capital moves more confidently. Markets reward alignment. Divergence introduces volatility.

Volatility is penalized.

Thus, engineered consensus becomes economically attractive.

This does not require centralized coordination. It requires distributed incentive compatibility. When alignment benefits all major actors, deviation becomes irrational.

Rational actors converge.

Convergence reduces variance.

Reduced variance increases systemic efficiency.

But efficiency can mask fragility.

When capital, policy, narrative, and synthetic systems reinforce each other, dissent does not disappear. It becomes economically disfavored. Funding narrows. Visibility drops. Institutional advancement slows for those who question structural premises.

The individual within such a system may remain intellectually independent — yet feel the cost of divergence.

Cost reshapes behavior.

Behavior shapes culture.

Culture narrows imagination.

Imagination narrows decision.

This is cognitive capture at scale.

The system does not demand agreement. It rewards alignment.

Reward is stronger than command.

The sovereign actor inside such a structure must recognize when institutional gravity influences interpretation. He must ask whether his perception is shaped by evidence or by reinforcement patterns that make alternative positions costly.

This is uncomfortable terrain.

Engineered reality does not merely shape what is seen. It shapes what is safe to see.

Safety is a powerful regulator of thought.

When safety aligns with infrastructure, the system stabilizes around a narrow band of acceptable interpretation. Leaders operating inside that band may appear rational and disciplined. They may in fact be so. Yet the boundaries within which they operate are externally reinforced.

The question is not whether consensus is wrong.

The question is whether consensus is tested.

Without deliberate testing, engineered reality solidifies into assumption.

Assumption becomes doctrine.

Doctrine becomes policy.

Policy becomes capital allocation.

Capital allocation reshapes infrastructure.

Infrastructure then amplifies the original assumption.

The loop closes.

Closed loops feel stable — until anomaly intrudes.

Engineered systems perform best under modeled conditions.

Models require assumptions. Assumptions require boundary conditions. Boundary conditions define what variables are included and which are excluded.

Within those parameters, performance can be extraordinary.

Speed increases.

Prediction improves.

Risk appears quantifiable.

Confidence expands.

But engineered systems inherit the constraints of their modeling assumptions. They do not fail because they are inaccurate. They fail because they are incomplete.

In stable periods, incompleteness is invisible.

Under stress, it is revealed.

Shock introduces variables that were deprioritized or excluded. Rare events. Human irrationality. Political acceleration. Institutional panic. Nonlinear escalation.

When these variables emerge, engineered systems do not immediately collapse. They hesitate.

Hesitation creates lag.

Lag introduces decision delay.

Delay amplifies uncertainty.

Uncertainty increases volatility.

Volatility disrupts consensus.

Consensus fragmentation destabilizes capital flow.

The feedback loop that once reinforced stability begins amplifying instability.

This is systemic brittleness.

Brittleness is not weakness. It is rigidity under pressure.

Systems optimized for smooth performance can become fragile when friction returns. The more friction has been engineered out, the less practiced institutions are at operating within it.

Engineered environments reduce exposure to edge conditions. Leaders become accustomed to curated information streams. Decision dashboards present distilled clarity. Anomalies are filtered as noise.

When a true anomaly appears, it resembles noise.

Noise is ignored.

Ignored anomalies compound.

Compounded anomalies become crisis.

Crisis reintroduces raw uncertainty into environments conditioned for compression. At this moment, the sovereign actor becomes critical.

If leadership is fully dependent on engineered summaries, it may overtrust model outputs precisely when models are least reliable. If leadership has cultivated dual vision — interpreting outputs while interrogating assumptions — it can widen perception before action narrows options.

This widening must occur quickly.

Engineered reality accelerates events. It compresses reaction time. Institutions conditioned to optimize for efficiency may struggle to expand cognitive aperture under pressure.

Efficiency resists expansion.

Expansion requires discomfort.

Discomfort requires discipline.

The sovereign actor must deliberately slow interpretation even when acceleration pressures him forward. He must reintroduce raw data exposure. Seek dissent. Invite contradiction. Test the model against lived conditions rather than projections.

This is not rejection of technology.

It is refusal to surrender judgment.

When shock enters a tightly aligned narrative-capital-policy system, cracks appear at the seams between assumptions. Institutions that previously moved in synchrony may diverge. Public messaging may fracture. Confidence may erode faster than underlying conditions justify.

Engineered consensus can unwind rapidly.

The danger is not collapse. The danger is overcorrection.

When engineered stability fails, institutions often swing toward reactive centralization or rigid control. Attempts to regain stability can narrow discourse further, penalize dissent more aggressively, and double down on model assumptions.

This increases brittleness.

The resilient alternative is adaptive sovereignty.

Adaptive sovereignty does not resist technology. It resists overidentification with it.

It recognizes that synthetic systems are tools, not epistemological authorities. It preserves independent verification channels. It maintains tolerance for dissent. It funds exploration outside dominant frames. It builds redundancy into interpretive systems.

Redundancy appears inefficient.

In stable periods, redundancy is questioned.

Under stress, redundancy is survival.

Engineered reality rewards compression. Sovereign resilience requires margin.

Margin in time.

Margin in capital.

Margin in interpretation.

Margin in institutional structure.

Without margin, systems become optimized but fragile.

With margin, systems appear slower but endure.

The temptation in engineered environments is to remove margin in pursuit of precision. Precision becomes identity. Speed becomes reputation. Optimization becomes culture.

But optimization without margin is performance without shock absorption.

The sovereign actor must ask: where have we removed friction that once revealed early warning signals?

Early warning signals often appear as inefficiency. They look like dissenting analysts. Outlier data points. Unpopular forecasts. Contrarian research. These are frequently deprioritized in optimized systems because they do not fit established predictive corridors.

Yet these signals are often the first indicators of structural stress.

Engineered reality does not eliminate early warning signals. It reclassifies them as statistical outliers.

Outliers matter.

The disciplined leader must cultivate an appetite for them.

Reasserting Sovereign Agency

Sovereignty in engineered reality is not withdrawal. It is composure under acceleration.

The sovereign actor does not attempt to dismantle infrastructure. He does not reject synthetic systems. He does not romanticize a pre-digital world that no longer exists.

He accepts the terrain.

Acceptance is the beginning of sovereignty.

Denial creates reaction. Reaction surrenders agency to tempo. Engineered environments are defined by tempo. Speed becomes structure. Urgency becomes culture. Response becomes reflex.

The sovereign actor refuses reflex.

Reflex is programmable.

Deliberation is not.

To reassert sovereign agency is to reintroduce disciplined pause into accelerated systems. It is to create internal latency where external systems compress time.

Latency is not delay. It is calibrated response.

Calibrated response requires layered perception. The disciplined mind must operate with dual vision: interpreting content while simultaneously evaluating context. Recognizing signal while interrogating ranking. Understanding narrative while identifying incentive.

Incentive is the hidden architecture.

Every engineered system has incentive alignment embedded within it. Ranking algorithms optimize engagement. Institutions optimize reputation. Capital optimizes return. Policy optimizes stability. Synthetic systems optimize predictive coherence.

Optimization is directional.

The sovereign actor asks: toward what?

When direction becomes invisible, alignment becomes unconscious.

Unconscious alignment is surrender.

The restoration of sovereignty begins with explicit awareness of directional bias within systems. Not to reject them — but to account for them.

Accounting creates margin.

Margin creates freedom.

Freedom is not the absence of structure. It is the ability to operate within structure without being absorbed by it.

Absorption occurs when identity fuses with infrastructure. When reputation depends entirely on alignment. When professional advancement requires interpretive conformity. When information intake is fully curated.

The sovereign actor cultivates independent channels.

Independent reading.

Independent data verification.

Independent capital relationships.

Independent interpretive networks.

These channels need not be oppositional. They must simply be plural.

Plurality prevents capture.

Capture thrives on singular dependency.

The engineered environment compresses plurality by rewarding efficiency. Multiple verification channels appear redundant. Diverse interpretive inputs appear slow. Dissent appears disruptive.

But plurality is resilience.

The sovereign actor therefore designs his own informational architecture. He chooses when to consume distilled summaries and when to access raw inputs. He separates model output from model assumption. He periodically disengages from optimized feeds to recalibrate perception.

Recalibration restores proportion.

Proportion tempers urgency.

Urgency is the fuel of engineered acceleration.

When urgency becomes constant, discernment declines.

The disciplined leader therefore distinguishes between systemic urgency and manufactured urgency. Not every acceleration demands response. Some demand observation.

Observation precedes decisive action.

In engineered reality, the most dangerous moment is often when action feels most necessary. Momentum builds. Consensus solidifies. Pressure intensifies. Visibility increases. Synthetic systems present high-confidence projections.

Confidence under compression can seduce even experienced leaders.

Sovereign agency requires comfort with standing still.

Standing still is not inaction. It is refusal to be moved prematurely.

Premature movement is exploitable.

Engineered systems, particularly those augmented by AI, can create cascading narrative momentum. Once a frame becomes dominant, alternative interpretations lose visibility rapidly. Leaders who move inside that cascade may later discover that underlying conditions were more complex than the frame allowed.

The sovereign actor resists cascades.

He widens aperture before narrowing commitment.

This widening may involve inviting dissenting analysis, commissioning independent modeling, revisiting first principles, or simply creating time between stimulus and response.

Time is sovereignty.

Compression reduces time. Reduced time reduces reflection. Reduced reflection increases model reliance.

Model reliance increases vulnerability to assumption error.

The restoration of sovereign agency is therefore inseparable from time discipline.

Time discipline means structuring decision environments to prevent reactive collapse. It means building institutions that reward thoughtful disagreement rather than penalize it. It means designing capital structures that tolerate variance. It means ensuring that synthetic systems inform rather than direct.

Direction must remain human.

This is not a romantic claim. It is a structural necessity. Synthetic systems optimize based on historical data and encoded objectives. They cannot perceive moral context beyond programmed constraints. They cannot evaluate long-term societal implications beyond modeled probability.

The sovereign actor integrates model output with ethical framework.

Ethical framework cannot be outsourced.

In engineered reality, the most subtle erosion of sovereignty occurs not when machines replace judgment, but when humans defer judgment because machine output appears more precise.

Precision is persuasive.

Persuasion can obscure responsibility.

Responsibility cannot be delegated.

The sovereign actor retains ownership of decision outcome even when relying on synthetic assistance. This ownership creates humility. Humility tempers overconfidence in predictive systems.

Humility is resilience.

Engineered environments reward assertiveness and speed. Sovereign agency requires quiet confidence and measured pace.

Measured pace is often invisible in public environments that reward immediacy. Yet durable institutions are built by leaders who understand that velocity without direction compounds error.

Direction must precede speed.

Speed without direction is amplification.

Amplification of error is collapse.

The sovereign leader therefore asks repeatedly: what assumptions underlie this projection? What incentives shape this narrative? What variables remain unmodeled? Where have we removed friction that once provided warning?

These questions are not obstruction. They are preservation.

Preservation of judgment.

Preservation of margin.

Preservation of agency.

Engineered reality is not a dystopia. It is an acceleration layer. It increases capability. It expands reach. It compresses time. It augments perception.

Augmentation is not sovereignty.

Sovereignty is disciplined integration of augmentation without surrender.

This integration requires internal architecture as deliberate as external systems. The leader must cultivate intellectual independence, ethical clarity, temporal discipline, and structural humility.

Without these, engineered systems will gradually shape him rather than assist him.

With them, engineered systems become force multipliers for sovereign judgment.

The future will not belong to those who reject synthetic augmentation.

Nor will it belong to those who submit to it unquestioningly.

It will belong to those who can stand inside acceleration without losing internal stillness.

Stillness is not passivity.

It is composure under compression.

Composure allows expansion when systems narrow. It allows plural input when consensus compresses. It allows ethical clarity when precision seduces.

This is the restoration of sovereign agency.

Not rebellion.

Not nostalgia.

Not resistance.

Integration with awareness.

Engineered reality is here. It will deepen. Synthetic systems will become more capable, more predictive, more embedded in institutional decision flows.

The sovereign actor does not fear this trajectory.

He prepares for it.

Preparation means cultivating cognitive plurality, institutional margin, ethical responsibility, and disciplined time.

In doing so, he transforms engineered reality from constraint into terrain.

Terrain does not dictate outcome.

Leadership does.

Institutional Architecture in Engineered Reality

Engineered reality does not merely influence individuals. It reshapes institutional behavior.

Institutions that once relied on deliberative pace now operate within accelerated cycles. Reporting intervals compress. Public response windows shrink. Predictive modeling replaces longitudinal judgment. Performance is measured in real-time metrics rather than generational durability.

The shift is subtle.

Institutions begin optimizing for visibility rather than resilience.

Visibility is rewarded in accelerated systems. Narrative dominance confers legitimacy. Data fluency signals competence. Algorithmic literacy becomes synonymous with strategic intelligence.

Yet institutions that optimize solely for engineered fluency risk losing structural depth.

Structural depth is slow. It requires historical memory, distributed authority, and tolerance for internal disagreement. It cannot be fully quantified. It resists compression.

Compression favors clarity over complexity.

Complexity, however, is reality.

When institutions reduce complexity in order to operate efficiently inside synthetic systems, they simplify internal debate. Dashboards replace dialogue. Predictive analytics substitute for dissenting interpretation. Decision trees supplant exploratory reasoning.

None of these tools are inherently harmful.

But when tools replace plural cognition, institutions narrow.

Narrow institutions respond well to expected variation. They struggle with structural rupture.

Structural rupture rarely announces itself through existing metrics. It emerges at the periphery. It appears first in areas considered marginal. It presents as low-probability noise.

Institutions conditioned to prioritize efficiency may ignore these signals.

Resilient institutions design for anomaly.

Designing for anomaly requires redundancy of interpretation. It requires multiple analytical teams with partially overlapping mandates. It requires preserving institutional memory that predates current modeling frameworks. It requires leadership that values friction as a diagnostic signal rather than a cultural defect.

Friction reveals stress.

Stress reveals weakness.

Weakness identified early is strength preserved.

In engineered environments, friction is often treated as inefficiency. It slows tempo. It complicates messaging. It disrupts predictive coherence.

But friction is often the first indicator that underlying assumptions are under strain.

Institutions that eliminate friction in pursuit of smooth operation eliminate early warning capacity.

This is where leadership matters most.

Leadership under synthetic pressure is tempted toward consolidation. Consolidation simplifies communication. It centralizes narrative control. It reduces interpretive variance.

In calm periods, consolidation appears disciplined.

Under shock, consolidation amplifies error.

Distributed interpretation increases resilience.

Resilience requires tolerance for partial disagreement.

Partial disagreement does not weaken institutions. It prevents blind consensus.

Blind consensus is brittle.

Brittleness under acceleration becomes collapse.

Engineered reality therefore presents a paradox: the very tools that increase institutional capability can increase institutional fragility if not integrated with structural humility.

Structural humility recognizes that no model captures totality. No algorithm anticipates every emergent behavior. No predictive system neutralizes uncertainty.

Institutions that internalize this humility remain adaptable.

Adaptability is slower than optimization.

But adaptability survives.

Engineered systems encourage performance metrics that reward short-term precision. Institutional survival requires long-term variance tolerance.

Variance tolerance appears inefficient in quarterly evaluation cycles. It becomes indispensable in decade-long arcs.

Leaders operating inside engineered environments must therefore design dual structures:

One optimized for performance.

One preserved for resilience.

The performance layer operates inside synthetic acceleration. It leverages predictive modeling, narrative infrastructure, and capital alignment.

The resilience layer preserves interpretive independence, historical continuity, dissent channels, and margin.

These layers must coexist.

If performance eliminates resilience, institutions become sleek and fragile.

If resilience ignores performance, institutions become principled but irrelevant.

Sovereign institutional design integrates both.

Integration requires conscious architecture.

Architecture requires intention.

Intention requires awareness of the terrain.

Engineered reality is not an enemy of institutions.

It is a stress multiplier.

Institutions that integrate synthetic systems without redesigning internal architecture risk amplifying their own blind spots.

Institutions that redesign consciously transform acceleration into advantage.

Consolidation: The Architecture of Engineered Reality

Engineered reality is not a conspiracy. It is convergence.

It emerges where narrative infrastructure, synthetic augmentation, capital alignment, institutional incentives, and cognitive optimization intersect.

Each element is rational in isolation.

Narrative infrastructure organizes meaning.

Synthetic systems increase predictive capacity.

Capital seeks stability.

Institutions pursue efficiency.

Individuals optimize for safety and advancement.

Convergence is natural.

What is unnatural is the speed.

Acceleration compresses feedback loops. When narrative, AI, capital, and institutional reinforcement begin operating at synchronized tempo, reality becomes shaped faster than individuals can recalibrate.

This is the condition we now inhabit.

The death of neutral reality was not the loss of objectivity. It was the fragmentation of shared interpretive frames. In response to fragmentation, synthetic systems stepped in to restore coherence through ranking and prediction.

Ranking is influence.

Prediction is direction.

Direction, when aligned with capital and institutional incentives, becomes structural force.

Force reshapes perception.

Perception reshapes decision.

Decision reshapes infrastructure.

Infrastructure reinforces narrative.

The loop stabilizes.

Inside this loop, individuals and institutions experience clarity — but clarity within defined boundaries.

Engineered reality is therefore not false reality. It is bounded reality.

Boundaries are not always visible.

Invisible boundaries are the risk.

The sovereign actor does not attempt to dismantle convergence. He maps it.

Mapping requires recognizing that every engineered system carries embedded incentives. Every narrative platform has directional bias. Every capital allocation reflects implicit assumptions about stability and growth.

Assumptions, when reinforced by synthetic coherence, become self-confirming.

Self-confirmation is comforting.

Comfort reduces vigilance.

Reduced vigilance increases capture.

Cognitive capture at scale does not require coercion. It requires reward alignment. When alignment produces professional safety, social affirmation, and capital advantage, divergence becomes costly.

Cost shapes behavior.

Behavior shapes culture.

Culture shapes possibility.

Possibility defines the perimeter of decision.

The sovereign actor must therefore cultivate awareness not merely of content but of perimeter.

Where does interpretation narrow?

Where does dissent lose oxygen?

Where does anomaly get reclassified as noise?

These questions preserve agency.

Engineered reality becomes fragile when its boundaries are mistaken for completeness. Institutions that confuse model performance with total understanding risk overextension. Leaders who substitute predictive confidence for structural humility risk acceleration without margin.

Margin is the quiet theme beneath all of this.

Margin in time prevents reactive collapse.

Margin in interpretation prevents blind consensus.

Margin in capital prevents forced alignment.

Margin in institutional design preserves resilience.

Without margin, systems appear precise but become brittle.

With margin, systems tolerate shock.

Shock is inevitable.

Acceleration guarantees it.

Engineered systems magnify both efficiency and error. When error propagates through synchronized narrative-capital-policy loops, correction can become destabilizing. Institutions that eliminated friction may struggle to reintroduce it under stress.

This is where sovereign agency becomes indispensable.

Sovereignty is not autonomy from systems. It is disciplined integration within them.

Sovereignty recognizes augmentation without surrendering judgment. It accepts synthetic capability without transferring ethical responsibility. It operates inside acceleration while preserving internal stillness.

Stillness is not passivity. It is structured composure.

Composure allows widening before narrowing. It permits dissent before decision. It encourages plural input before capital deployment. It invites recalibration before reaction.

In engineered reality, tempo is power.

The sovereign actor manages tempo rather than being managed by it.

This requires dual structures — personal and institutional.

At the personal level, sovereignty demands independent verification channels, deliberate exposure to anomaly, disciplined time management, and ethical clarity beyond model output.

At the institutional level, sovereignty demands architectural humility. Distributed interpretation. Redundant analysis. Incentive structures that tolerate dissent. Capital frameworks that preserve margin.

Engineered reality will not slow.

Synthetic systems will grow more predictive. Narrative infrastructure will become more refined. Capital will move faster. Institutional feedback loops will tighten.

The question is not whether this trajectory continues.

The question is whether those operating inside it retain authorship.

Authorship is the final distinction between sovereignty and absorption.

Absorption occurs when direction is inherited unconsciously from infrastructure.

Authorship occurs when direction is chosen consciously within infrastructure.

The sovereign actor reclaims authorship by interrogating assumptions, preserving margin, and integrating augmentation without dependency.

Engineered reality is terrain.

Terrain does not command.

It conditions.

Leadership determines whether conditions become constraints or advantages.

This volume extends the architecture established in Sovereign Intelligence. If the first examined the disciplined mind inside complexity, this examines the environment shaping that mind.

Sovereign intelligence was internal architecture.

Engineered reality is external architecture.

The two are inseparable.

Without sovereign intelligence, engineered reality shapes the actor.

With sovereign intelligence, engineered reality becomes navigable.

The future will not be decided by those who reject acceleration, nor by those who surrender to it.

It will be shaped by those who understand that technology is an amplifier, not an author.

Authorship remains human.

Responsibility remains human.

Direction remains human.

Engineered systems may rank information. They may model probability. They may simulate coherence.

They do not own consequence.

Consequence returns to leadership.

Leadership, in accelerated environments, must be quieter than the systems it commands.

Quiet does not mean weak.

It means grounded.

Grounded leaders preserve internal sovereignty while operating within engineered systems. They build institutions with margin. They cultivate dissent without chaos. They integrate synthetic tools without deferring moral agency.

They understand that precision is powerful, but humility is stabilizing.

Engineered reality is not destiny.

It is design.

Design can be understood.

What can be understood can be navigated.

Navigation is sovereignty.

Final Reinforcement: Design, Agency, and the Future of Direction

The modern environment is not chaotic. It is structured.

Its structure is layered:

Narrative infrastructure organizes meaning.

Synthetic systems accelerate interpretation.

Capital aligns incentive.

Institutions compress decision cycles.

Individuals optimize within reward systems.

These layers now operate in synchronized tempo.

That synchronization is the defining feature of engineered reality.

When tempo aligns across systems, direction emerges quickly. Narratives gain dominance rapidly. Capital reallocates with speed. Institutions pivot in compressed intervals. Public consensus forms or fractures within days rather than years.

Acceleration increases influence.

Influence shapes perception.

Perception shapes possibility.

Possibility defines strategy.

This is why engineered reality must be understood as architecture rather than atmosphere. It is built from systems, incentives, and reinforcement loops. It is designed, even when not centrally coordinated.

Design creates corridors.

Corridors are efficient.

But corridors are not infinite.

When leaders mistake corridors for total landscape, they narrow their own strategic field.

The first movement of this volume established that neutral reality has dissolved into ranked visibility. What is seen is curated. What is amplified is optimized. What is repeated becomes stable.

The second movement examined narrative infrastructure — the scaffolding through which meaning travels. Language, framing, repetition, and visibility define interpretive dominance.

The third movement addressed synthetic augmentation. Artificial systems now shape probability assessment, signal prioritization, and predictive modeling at scale. Augmentation increases capability but introduces assumption inheritance.

The fourth movement exposed cognitive capture — not as ideological possession but as structural narrowing reinforced by incentive alignment.

The fifth movement analyzed institutional and capital convergence, where efficiency, stability, and reputational management reinforce engineered consensus.

Together, these five arcs describe the environment.

But description is not destiny.

Destiny emerges when environment goes unquestioned.

Engineered reality becomes constraining only when it is mistaken for inevitability.

Inevitability is the final illusion.

Systems accelerate. They do not predetermine.

Acceleration compresses time, but it does not eliminate agency.

Agency contracts when individuals and institutions surrender authorship to system outputs.

Authorship is preserved through conscious integration.

Conscious integration requires three disciplines:

First, interpretive discipline.

The leader must distinguish between ranked signal and raw signal. He must ask not only what is visible, but why it is visible.

Second, structural discipline.

Institutions must design for anomaly and dissent, not merely optimize for smooth performance.

Third, temporal discipline.

Decision cycles must include deliberate pause even inside accelerated environments.

These disciplines are not intuitive under synthetic pressure. They must be cultivated.

Cultivation implies intention.

Intention implies awareness.

Awareness begins with naming the architecture.

That is the purpose of this volume.

If Sovereign Intelligence examined the disciplined mind, Engineered Reality examines the disciplined environment.

The former focused inward.

The latter focuses outward.

The inward architecture preserves clarity.

The outward architecture shapes context.

Without inward clarity, outward forces dominate.

Without outward awareness, inward clarity is misapplied.

The two volumes are not sequential arguments. They are complementary structures.

Sovereign intelligence provides the internal compass.

Engineered reality maps the terrain.

Compass without terrain produces misdirection.

Terrain without compass produces drift.

The future belongs to leaders who integrate both.

Such leaders will not appear reactive. They will not appear frantic. They will not attempt to dominate narrative through volume alone.

They will move with measured pace inside accelerated systems.

They will question assumption without theatrical opposition.

They will integrate synthetic systems without surrendering ethical authorship.

They will build institutions that perform and endure.

Engineered reality will deepen. Predictive systems will grow more refined. Ranking architectures will become more subtle. Capital will move faster still.

But direction remains a human act.

Technology can amplify.

It cannot decide meaning.

Meaning remains authored.

Authorship remains accountable.

Accountability remains human.

The question facing leadership is therefore not whether engineered systems will continue to expand.

They will.

The question is whether leaders will cultivate the margin, plurality, humility, and temporal discipline required to remain sovereign inside them.

Engineered reality is not an adversary.

It is an amplifier.

Amplifiers increase both signal and error.

Without disciplined integration, amplification destabilizes.

With disciplined integration, amplification empowers.

The distinction is not technical.

It is architectural.

Architecture is chosen.

Choice is sovereignty.

And sovereignty, properly understood, is not isolation from systems.

It is authorship within them.

Coda: The Discipline of Continuity

Every technological age believes it stands at the edge of transformation.

What distinguishes this era is not transformation alone, but continuity under acceleration. Systems now adjust in real time. Narratives recalibrate within hours. Capital reallocates within minutes. Synthetic systems learn continuously.

Continuity under acceleration produces a subtle illusion: that change itself becomes the organizing principle.

But change is not the principle.

Design is.

Engineered reality is not perpetual upheaval. It is structured adjustment. Ranking shifts. Models refine. Incentives realign. Visibility reorganizes. The environment feels fluid, but the architecture remains.

The architecture is composed of incentives, reinforcement loops, synthetic augmentation, institutional compression, and capital alignment.

Understanding this architecture restores proportion.

Proportion restores discipline.

Discipline restores sovereignty.

When leaders view engineered reality as chaos, they react. When they view it as design, they navigate.

Navigation requires steadiness.

Steadiness requires internal coherence that does not fluctuate with narrative velocity.

Velocity will increase.

Synthetic systems will predict more accurately. Institutional cycles will compress further. Public reaction windows will shrink. The pressure to move quickly will intensify.

But sovereignty is not speed.

It is continuity of judgment across changing conditions.

Continuity requires principles that predate infrastructure.

It requires leaders who understand that technology amplifies direction but does not determine it. That capital reinforces assumption but does not author it. That narrative shapes perception but does not eliminate consequence.

Consequence persists beyond models.

Models update.

Consequence remains.

Engineered reality will continue to evolve. Synthetic sovereignty will tempt institutions to substitute prediction for prudence. Optimization will reward alignment. Acceleration will reward visibility.

But disciplined leadership will reward margin.

Margin is not excess.

It is space for recalibration.

Recalibration prevents collapse.

Collapse is rarely immediate. It begins with overconfidence in bounded systems. It accelerates when feedback loops go unexamined. It stabilizes only when humility reenters architecture.

Humility is not weakness. It is structural awareness of incompleteness. Incomplete systems can function extraordinarily well — until they encounter variables they did not anticipate. Sovereign leaders anticipate incompleteness. They preserve dissent. They cultivate independent verification. They build institutions that can tolerate friction. They move deliberately when acceleration tempts reaction. They integrate synthetic systems without surrendering authorship. They remember that design can be understood. What can be understood can be shaped. What can be shaped can be navigated. Engineered reality is not fate.

It is terrain.

And terrain, however complex, remains navigable to those who retain compass and composure.

Closing Note

Engineered environments do not announce themselves.

They operate quietly. They optimize invisibly. They adjust incrementally. Over time, their influence becomes ambient — less noticed precisely because it is everywhere.

The disciplined mind does not resist architecture. It studies it.

Throughout history, power has shifted toward those who understood structure rather than surface. In earlier eras, that structure was territorial, industrial, financial, or institutional. In this era, structure is increasingly informational and computational.

The question is no longer whether reality is mediated. It is.

The question is whether those who move within engineered environments understand the rules embedded within them.

Clarity remains a sovereign act.

Strategy remains a structural discipline.

Agency remains possible. But none of these endure without awareness of architecture. To navigate engineered reality is not to withdraw from it. It is to see it clearly enough to operate within it without surrendering autonomy.

That work is ongoing.

About the Author

Joe Cozart is the founder of GMJoe™ Consulting, where his brand anchor — Clarity. Strategy. Sovereignty. — guides his work across strategic analysis, institutional architecture, and complex decision environments.

His writing explores the structural forces that shape power, perception, and agency in modern systems. Blending disciplined analysis with cultural intelligence, his work examines how engineered environments influence leadership, narrative formation, and institutional resilience.

Engineered Reality continues his examination of sovereignty in an era defined by algorithmic mediation and designed informational ecosystems.

Made in the USA
Monee, IL
04 November 2026

67387306R10059